The Plan to Save all Life on Earth

ISBN: 978-0-57864838-5

Cover photo- Arthur Drew Sidell

Cataloguing and publication data

Duffy III, William Charles, 1970-

The Plan to Save all Life on Earth/ William Charles Duffy III

Includes Table of Contents

ISBN 978-0-57864838-5

1. Science – Environmental 2. Teaching – Science

The Plan to Save all Life on Earth

by

William Charles Duffy III

GOING to press, it's probably true: "Temperature and Electromagnetic energy denote the same thing, which is wave energy." They distinguish themselves, however, by (1.) working only within the other's effective environment, to separately affect Life of Earth and (2.) dissipating to (or approaching) *exclusion of relevance* separately, i.e. "only" through extraordinary distance of air, "atmosphere", or space (electro) and "only" through contained mass (temperature). This second point regarding temperature is not intuitive, mind you, until you define the living *environments* of Earth as the full range of subject matter. *It* may similarly be observed that Mass and Gas denote the same thing, which is matter, though the two are separate in their affect on Earth because one is suspended *above or within* all matter otherwise bound by gravity to the mass of Earth (gas) and one is part of a whole of singular composition bound as closely as molecularly possible by gravity (mass). (It is, indeed, more so important to delineate liquid as mass, which one will see later in this work.)

This work is written with the knowledge that since the last *near extinction of life on Earth*, by ice age or meteor, there hasn't been one. Therefore, the aspects described herein as being to the detriment of the habitability of Earth, without exception, are the fault of mankind's encroachment upon it (and its close environs.) It is not my intent to exclude readership of this text. However, one so wrong or conditioned to assume that the progression of problems on Earth in the last few thousand years, increasing, as you're aware, by exponent, can be considered natural (conveniently, to the point it separates humanity from blame) is even mathematically possible considering the length of tenure of the planet and the various life forms it has supported in that span may omit his or her need to progress further into this document.

Drought   Skin Cancer   Floods   Desalinization by Fresh Water Displacement of the Oceans   Altered PH of the Oceans

Tsunami   Gun Violence   Car Wrecks   Leukemia   Declining Bee Population   Loss of Mass/ Base Weight on Earth

Irrevocable Loss of Natural Resources   Cancer by Inhalant Toxins   Global Temperature Extremes   Abnormal Hailstorms

Extinction of Animal Species   Greenhouse Effect   Localized Temperature Increases   Wildfires   Haboobs   Solar Flares

Abnormal High/ Low Weather Pressure Systems   Altered Chemical Barriers to Space   Depletion of Solar Radiant Energy

Lack of Education v.s. Technological Ability   Hopelessness   Fear   False Definitions of Global Warming   Crop Loss

Altered Celestial Relationship by Distance, Speed or Angle   General Disease   Increased Harmful Microorganisms

---

Mass"Displacement",Global"Warming",ElectromagneticEnergy"Displacement","Atmospheric"Gas(es),"Over"population

Recycling (Need)   **Caustic Waste (Do not need)**

Immediately above is the "Plan" Outline, which repels/ supports/ contains/ or represses all known Environmental, Ecological and Astronomical threats to Life on Earth. It is not the "General Outline of Science" which is contained in paragraph 6, below.

Introduction

It is the intent of the creator of this address that the basic outline provided be used to augment educational curricula for students internationally, immediately and without exception, as a basis with which they and others may more completely understand, examine and explain the environmental, ecological and astronomic threats which face life on this planet. Those threats to which all others are inexorably linked are enclosed here. They are, as above, the following: Mass Displacement on Earth, Global "Warming," Electromagnetic Energy Displacement on Earth, Atmospheric Gas Displacement on Earth (Ozone, etc.) and Overpopulation. Also, as above, but less thoroughly covered in this volume: The need for Caustic Waste *Disposal* and *Recycling of reusable* products. It is a more in- depth explanation of the first five of these items than that which might be presented to children in schools (in an outline form) contained in this book.

It is the belief of the writer that this is the scholarly outline to preclude the tasks ahead of us, and that deciphering the outline as it may be required, to ingest and change it to an appropriate format for educators, provided that its core elements and their parts are included, is a task just fine for others.

It is also the hope that actual works specified in this address be employed now, to the extent that they can be. Work to analyze and restore the atmosphere, calculate mass and electromagnetic energy displacement, and work to collect proper data on other outline points, may indeed begin immediately.

This thesis, written, may be added to any tidbits found exclusively in an attached videography project hosted by the writer, or vice versa should it allow a better text (a.) or scrolling by viewers who choose to see it online or on television (b.).

Nothing omitted here- known as fact to various types of scientists- is refuted by this address or the accompanying movie if it is not mentioned. Considering the movie, by example, chemicals as impediments to bees, is valid, for instance, and brakes on cars may as much be to blame for asbestos in the air as they are, etc.

An appropriate explanation of Global Warming is provided below, someplace. This explanation shows the phenomenon *sharing blame* for the altered natural weather pattern and ecological state of the planet with other threat outline components. Previous attempts to explain it were incomplete and ineffective. Work to counter this phenomenon under circumstances which exclude the proper elements or exclude the necessary ones will fail.

With all respect, the "general outline of science", as opposed to the threat outline to which this item is dedicated, is: Mass, Temperature, Electromagnetic Energy, Gas(es) and what is affectionately known as "the Multiple," the constant, meaning people or the population count, and the result, without bias, negative or positive, by them upon the Earth and its surrounding systems. (Although we will see this more clearly as we progress in this work, it is not the point of it, and so... for that second Nobel prize you may count my humility as a point scored in favor.)

It is or was a "conundrum" of my work, the worst of them, that, *as I progressed*, I would not be able to fill in the blanks fast enough for those who might decide for themselves- burn into their consciousness, as it were- what threats faced humanity and other life on Earth with the most severity (based on the "general outline of science," first written in this same paragraph, above.) Concurrently, the question surfaced as to whether or not there would be those who felt something was missing from my efforts for the omission of it. As I progressed with no mention of the "outline of general science", as it may otherwise be called, which I had inadvertently provided for humanity, it seemed that this was indeed the case. My decision was to keep the explanation herein, (albeit limited to the few paragraphs you are in the midst of now.)

This is a *measurable*; E = Electromagnetic energy (displacement) **and** Temperature (warming), M = Mass (displacement) **and** (atmospheric) Gases (particularly Ozone) and C = Population (Over) in order to reach E=MC2.

As we look at "threats to life on Earth", we come to a need to admit that the factitious ("man- made") affect upon the Earth is the entire OR chief cause of the threats to life on Earth which affect that group: the human Population and other Life on Earth. Therefore, we come for the other items listed to the factitious/man-made affect upon them *as well as* the affect upon them which humanity could or can do nothing about- as they occurred or may, either one- and as they themselves affect said Life on Earth itself.

This address is entirely dedicated to an explanation of how an outline of scientific information based on the itemization of scientific threats (based on the components of science itself) completely repels these threats. Thereby, these threats are, for your Mass, Mass "Displacement" on Earth, for your Temperature, "Global Warming", for your Electromagnetic Energy, Electromagnetic Energy "Displacement" on the Earth ("to" and "from" each included), and for your Gases (predominantly, as explained further on), your "Atmospheric" Gases, depletion of Ozone in particular. It is, for your Population, "Overpopulation." Bookends, as they are, to that threat outline, again, as we consider life on Earth, those things we don't want, those that may kill us, we call "Caustic Waste", and those things we need- unwarrantedly disposed of to the point of inconvenient or *possibly impossible* retrieval- the “need for Recycling.” That becomes your “threat” outline, as described firstly in the diagram at the top of this document, and the outline of information described in the accompanying movie on the subject matter. That’s your level. That’s your level of information which must be freely shared to international curriculums.

Atmospheric &

Earth bound Gas

Mass (Ozone & C02)

Electromagnetic Global \ / Over-

Energy Displacement Warming Displacement Population

\ / \/ /

E = M C 2

/ \ / \ \

Electromagnetic Temperature Mass Gas Population

Energy

Unfortunately, it’s been quite some time since I wrote the first draft of this item. I doubt that people who knew me thought I had such considerations. It was prior to this- perhaps accurately explained as “over most of my life”- that the elements in it, those of severity, stacked themselves in my consciousness. They became subdivided properly by their parts, and each then related to each other. My understanding of these items came as a person’s understanding of anything does, over time. It was, as I remember it, reading the paper with my family at breakfast before school as a boy that warming trends on Earth and the depletion of the Ozone layer were revealed to me as issues of some concern to scientists and educators. This was in the 1970s or perhaps the early ‘80s. These things weren’t and aren’t what people enjoy discussing, in any case, and so I kept them to myself unless prompted to otherwise, generally in school. I wrote this thesis based on my understanding of the problems we faced as people on Earth. I rose toward the end of my typing and stared at the wall for a moment, my gaze locked on a point the least remarkable of my mother’s home, a wall, a closet door, an odd desk corner and end table... It was my thought at the time that the address was not in a form to be ingested by humanity, and that I wasn’t well heeled enough to make it such that it would be; I opened the closet door and secreted the document into my filing cabinet. Years later, I pulled the item from that same box and reviewed it. I had done so only a few times before. Nothing that the thesis bore relevance to had changed, and only worsened, and as my memory of what I had written was correct and corresponded still to the type-written paragraphs below.

In any case, it was with the understanding by me that all other options (not this one which you read) would not be good for life on Earth, and that this “plan outline” could not be equaled, I elected myself, after a long consideration- or was so by someone else if you choose not to imagine me so powerful- to the task of assessment of our factitious, *or man- made*, ecological reality. That the item related well enough to the possible effects of non-man-made threats was a bonus. It does. These threats related just as well with the outline of information provided; indeed, they found themselves confused with such items to the point of irrelevant obscurity as I sat down to type.

I am, as they say, a ringer, a shark, or a specialist. I have succeeded in the equation relevant to what is most important. If you allow me leeway, now, to refrain from explanation of this insistence, except by example which I will provide now, I would appreciate it: In my work, a *choice* I had to make in providing you with the worst of the worst and how to stop it involved *atmospheric depletion*. There were many choices I had to make in this address and this one was no different. It provides this example as well as any. Your threat assessment relevant to "gases" on what would otherwise be an address strictly adherent to the Outline of General Science- or scientific elements which lead to the assumption of threats- would leave scholars and educators an *argument.* I'm not a big fan of leaving you in a quandary. My biggest point, if there is one, the reasoning for this whole address, is that humanity shares a common threat outline with each other, comfortable at least in receipt of it that the next scientist, in the next town or country, has the same information. It became then that the possibility of years of disagreement between scholars might waste time in the application of effort to alleviate these threats that I acted and this case of threats relevant to "gases" left the possibility that gases underground might be the threat to look at.

Nightmarishly, should work to finance the items herein begin, and a number of those in receipt of this work did not spend such sufferings over time as I endured to logically figure out which threats to address, the possibility a scholar might stall positive effort by pushing that the money and effort allocated to the tasks of atmospheric work be put toward the increased draining of pressurized underground gases on a quest for relevant heat and pressure instead of effort to alleviate the situation of the atmosphere, the diversionary effort would decrease the time that Life might exist on this planet. It may not be the last time you infer a certain sentiment from me as we progress- particularly with respect to "global cooling," which is touched on later- but the end result of threats to the *atmosphere* should we not attend to it are a far higher probability of indicating a global extinction than any possibility resulting that humanity should drill holes into the earth to forestall the eruption of a super volcano. A superheated exocrust due to global warming, and ignoring it, is what will lead to this secondary requirement if it ever gets that far.

This work, and *any* put forth on behalf of humanity, may eventually be accomplished as required without the argument, and without the omission of those items which are absolutely required. In this case that gaseous requirement is the atmosphere. At this time, very thankfully, there is no other world-wide effort required to be attended to relevant to "gases" and, really, though a volcano may do so, absolutely, this *planet* is not more likely to explode as the reason for its demise than to both burn and freeze yearly going forward.

I will assume that completion of this address by a reader shall accommodate any defense of it. My history, as mentioned, includes that I prepared myself or was so during my life for the task, to the point you read now, at any rate. The result for those who endeavor to know, to teach, etc., is the same if you can slog through the rest of it.

The Plan to Save all Life on Earth by William Charles Duffy III

People are likely to process and act on information differently than in the age before television and computers, movies and radio, cars and cubicles. People have been socially programmed to avoid interaction with others, and, specifically, to disagree with each other in areas of Earth dominated by such advancements. This is relative to all subjects which they have not put themselves, specifically, to the task of learning from another, or- by membership, generally, to any number of "societal systems" to which they are conscripted- resigned to agree with. Difficult, then, for an absolute supra genius to rationalize his undertaking the most important work in history.

Additionally, any amount of humor that you see in this work comes from I know not where... perhaps a laugh is proof of my retention through the adventure of my life of my own humanity. I would beseech you not to hold against me as I do my work the severity of tone with which I write or speak. Earth's ecological threats are 7. All others are related to them. 5 core elements that work together on behalf of each other, in concert, to the detriment of life on Earth are detailed below: detailed, presumably, to the benefit of the existence of all life. Two additional elements, those things we need, (recycling or lack thereof) and those we don't (caustic waste) are cursorily touched on later.

If it wasn't the intent of this address to maximize the tenure of all life on Earth, and only to lay out the problems it faces, then I would care a little less that one didn't believe me or did so, *applying myself, as it would be then (if the expectation was that one didn't) to membership of some group, perhaps, "against" the uninformed, etc. It is, however. These words alone, with my humility, represent the longest possible tenure and extension of tenure of all life on Earth. Humanity cannot do better for itself than to ascribe in schools to the proliferation of curriculum that teaches the common outline, pretentiously included, whose specific components may, for proof, alone, be plugged in as variables to the baseline of science, the formula E=MC squared. An assured result of action can come from no other method.*

Existentially, many of my blocks to you, and yours to me- without insult they are your "cowardices"- if we agree, are now gone. These are, conventionally, known as "psychological defense mechanisms" and the two relate to beliefs which are not based "entirely" on reality, but the fear instilled by confronting reality which the mass of people in our society, those around us, have chosen not to. Though it is not relevant to your ability to see truth through your psychological blocks- and therefore not relevant to your ability to ingest what I have to say- a note, now, on the progression of existential philosophy: there is no effective "post-existentialism" that I have ever encountered. The only logical progression of it is that the philosophy may include plants and animals as living beings sharing this planet, our experiences, and our lives, and, if I may, until proven otherwise, the Sun, especially, and other antiques of our solar system, as well.

Concerning that, my insistence, this- as close to my life's work as any may assumed to be- is based on the total of my very general knowledge of life as I see it, assisted by professors, and media, for the good of it, from time to time. My insistence is that it is possible for me, and for society, by a certain understanding, to accommodate everybody. This is my personal best, assimilated due to a variety of circumstances, seemingly improbable at times, and existential fortitude.

We go now to the sections of this work that define and then solve different serious problems facing the only mass of life in the universe which we will ever know, our world. We leave behind what the rationale might be to propagate a system that keeps people psychologically wary and geographically separate from each other, save for the following reiteration- which may be more easily accepted if you have lived as I, without a civil war since the last one: The system which keeps people from affecting positive change in their world **also** does not stand to allow humans to instigate negative change, nor to allow the system itself to fall, for any type of uprising the likes of which history, recorded, has shown very much. Such a system resists organization and therefore stands less the chance to see takeover by fascists or Marxists, a race- based genocide, a holocaust, or the great, armed, general purges that mark human history.

OVERPOPULATION

Overpopulation is the *outline component* we cover first because it is the issue (relevant to both space and consumption) which contributes to the others. (It confronts us *as us*, if you will, and *is the reason* for the others, as we have mentioned someplace above.)

Until now, overpopulation was the most difficult to deal with in an ethical way. If you are a power of some sort that is considering dealing with it, solving it, then perhaps the easy, ethical way *is* how it becomes the *easiest* of the outline elements to deal with, (rather than the most difficult or painful.)

The population timeline curve today is distinct. As I understand it, the birthrate, worldwide, has ebbed slightly since the onset *of technology,* and may continue to do so. *This specific cause and effect, lacking a specific definition, is therefore localized on Earth,* and leaves us with something else: that for a different scholar.

Those items described at the outset of this address have lain their effect on the birthrate, and we're all quite sure- in consideration of those items only- that somebody should be congratulating themselves for the accommodation. However, now, the birthrate is such that the eight or so billion people who we must provide for in this world will be eleven billion in forty or fifty years, and there will be *less* land and food then. A list of nasty items, just a reminder, including HIV/Aids, Cancer, Overdoses, Heart Disease, Car Wrecks and War… don't seem to be doing much to forestall the imbalance.

This reveals that the tremendous efforts made by those intending to kill others- or those whose efforts or lack thereof create the same circumstance by *unintended* consequence- result in failure, save to cause an expedited doom to everybody via a normalization of the psychology of genocide. Therefore, it is most important to take intentional, non-destructive steps to curtail the birthrate. It is that we live now, at this time in our progression as a species- and most certainly not for the reason which is that I write this- we are capable. We simply are.

Specifically, what we face, if you'll suffice it to say, at the original writing of the draft of this work, eight billion people in consideration were seven and an edit was required to change the digit.

Unfortunately, as time went by, in the western world, foolish men and women were heeded more than they should have been and those people carried a voice of such ignorance that the order of the day, the argument put forth, was for more people, and an increased birthrate!

The four steps listed below are relevant to two issues; the first three to childbirth, and the last to human rights concerning the demographic on earth who may not like to have children in the first place.

A precursor to my best salutations to humanity at the end of this work is that these items should not be forced upon populations, but seen as an understanding by nations and educational systems as only the best they can do for their people and the people of Earth.

1. *Early term abortion is to be made legal in all nations of the world. It is not suggested here that people are forced to abort children, and this line is typed so that you know it. The first step is proscribed that such abortions be made legal, provided the woman can make a free and informed, educated decision concerning it, and is given a safe facility to attend in the case of the decision.*

2. *Healthcare systems of all types, including non-socialized healthcare systems, are to be made to subsidize the small percent of their business that is pertinent to birth control. This means, at the least, no health clinic shall be forced by law to deny, for example, condoms, birth control pills, or, as important, "day after birth control" pills to those who ask for them. Regardless of nation, for every health clinic worldwide, this system must be socialized or "attended to" if you seek a different method. Socialized, herein, means, again, subsidized somehow where it is required.*

An unfortunate note on unethical exploitation of such a system: in the potential case of birth control used by sexual slavers to assist in the propagation of their trade, a mind during the implementation of such programs should be paid toward this possibility. Though the slaver and the patient may both be in favor of the result, the patient is who matters. As it is now, the opportunity to assist or free such a victim is at this confluence of forced procreation and the legal attempt to delete the pregnancy. Should such systems be expanded, those policy items concerning a patient's social services and criminal's legal responsibilities are so as well.

3\. *Educational systems augmented where they are not, currently, worldwide, to include this address, shall also include health education classes which explain birth, birth control, health education and, specifically, "sexual identity as a personal decision", each.*

This system must understand and relay to students the simple, actual statistics of overpopulation. This shared understanding is considered most important, whereby people may make informed decisions about family planning.

4\. *As legal and educational issues may differ, laws condemning homosexuality must be repealed, or else the nation in question should be sanctioned under international laws pertaining to human rights abuse. In terms of education, this method provides for both children and adult antagonists to the free lifestyles of humanity.*

That a demographic exists preferring a same sex lifestyle and an occasional desire to have children, or adopt them, is immaterial, and not unethical. The worldwide, legal legitimization of homosexuality is the strongest recommendation. In addition to being contributory to the goal of lowering the world's birth rate, also, existentially, this stands as the legitimization of the personal individual and their abilities to make decisions for themselves; thereby, you have a definitive representation of freedom for which all people strive.

It is quite possible for men and women of any land to express tolerance and understanding- as much so, anyway, as their nation's history shows them having acted the opposite.

In closing, we reference Item 3, above, and our suggestion that we approach educational systems to teach the rate of overpopulation. As the "rate of overpopulation" becomes an integral part of curriculums for children, the advancing population of Earth will begin, only a few years after, to ease in its growth due to the Earth's populations' understanding of it, collectively, and for the first time in history. Responsible family planning, not from law, but truthfully obtained statistics, explained as such to a young demographic, and in an educational environment, is the order.

It is from this ethical point, one of respect for other people, that the problem becomes solved, *therefore*, in a way whereby we become socially advanced in our societies. We also become truly advanced in our understanding of, and contributory to, the emotional and social well-being of the planet; the economies of nations become less encumbered by moral travesty.

After a short time of effort by the nations of Earth these steps may be implemented, and to the lowering of the explosive birthrate on Earth. The steps exclude, obviously, purposefully, "something(s) being done about it", including those "things" written above *and not so*, as evidenced by history, for reasons which are clear: We shall exclude other options which are unethical.

GLOBAL WARMING

In this chapter we endeavor to explain global warming. The explanation has no precedent, or none of its type. That there is no better solution, and because I say so, does not require that you search for one. In humility, the reason for this address is to share it with all people and that required modifications to it, if any, may be done together.

Explanation of the *Greenhouse effect, Ambient heat* and *Deforestation* follow as they constitute Global Warming's entirety.

Global Warming, "**the Greenhouse effect**."

Due in general to the large population's use of cars, and all the cows and chickens we farm to eat, etc., there is an increase, daily, in the amount of Carbon Dioxide, and other *greenhouse gases*, Methane included, released into the atmosphere. Without a detailed explanation, other heat trapping chemicals than CO2- also known as greenhouse gases- act to trap heat on Earth just as you are aware that CO2 does.

Animals and humans breathe in to get Oxygen, mostly. They breathe out to expel Carbon Dioxide, or CO2. With this increase of CO2 provided by humans and animals to the surrounding environment, there is also an increase of CO2 emitted by the machines that humans create and use to do things like carry them around. Cars excrete a large amount of CO2, and so do planes and smokestacks and lawnmowers, etc. What CO2 does when it is released is fill up the atmosphere, up to the very levels which separate our planet from outer space.

The Sun creates warmth on Earth. It sends its rays to Earth from many thousands of miles away. When they get here, the rays, they hit us, and everything, and "warm us up", so we're not cold like space. After the rays hit earth, they bounce right up off us, and take their heat with them, back out into space. This has been the way for millions of years on this planet. However, we see something interesting has happened. In the past *century, or few centuries*, humans have seen to it that more CO2 be released than ever before. That CO2 has created, and is perpetually creating, a layer, a *chemical ceiling*. The sun's heat now stays on Earth, trapped by that layer of CO2. Where it was once capable of rising and dissipating at the cooling levels of space itself, after it had done its job on Earth, warming us, the Sun's heat (and other heat that we will explain) now stays on Earth longer, trapped. The effect, this "greenhouse effect" as it is called, as you may know, makes the surface temperature on Earth get incrementally higher every year, around the world. It is also, partially, responsible for penguins and polar bears losing their homes into the oceans.

But that is one component of global warming. It is the most prevalent, the most explained. A student of ecological distress cannot be without a cursory understanding of this phenomenon.

Global Warming, "**Ambient Heat**."

We begin by posing to ourselves a question: How many cars would you guess there are driving around right now, on Earth? I've decided that 500 million is a guess from which an idea of reality and an easy enough computation may both be made. That's about half of the gas burning cars that are operational on Earth. Now, if the amount of CO2 coming out of one car is more than the amount coming out of no cars, and you can multiply that amount by 500 million, then you have that amount of CO2 released by cars and nothing else. The computation is simple enough, and it has been done, but the amount is not put forth on my browser window or the evening news, and so it is. The point is moot. An increase in global temperature might very well be caused by excess CO2 trapping the naturally occurring heat of the sun on Earth, but an equivalent problem, possibly *most* of the problem, is that there is an increase in temperature created on Earth by the machines made by and for humans. That is to say, the heat *we create-* not the heat trapped on Earth which was originally in possession of the sun- *is the predominant cause of increased temperature*. That amount may be computed using the exact same method as we do CO2, and thus we hypothesize:

If you take your car to a shop to be repaired, and the problem cannot be accessed under the hood or bonnet, then the mechanic must usually raise the car up on jack stands or a lift. If your mechanic is nice, or your brother-in-law, he may let you stay in the garage with him while he examines your car's underside, and then assess whatever problem your car has, and show it to you, explain it, in person. While he's telling you about your shocks or brakes, you might lean on the most accessible part of your car, for weariness or stupidity, and fuse your hand to the tailpipe where it will turn black and red and purple, and from whence will come a hiss and smoke that smells of human flesh, while you scream in terror and pain. Why? Because the exhaust manifold of a running car, or recently running car- from the underside of the motor (all of which is hot), and all the way to the end of the pipes sticking out of the back of the car, equal to 400 degrees(ish) F®, again, for argument, and the figure in use for our computation.

So, now burned, you calculate for yourself the number of cars emanating 400-degree temperatures while a foot or so over surfaces of cement *which retain heat quite well*. You use the figure 500 million cars, and then multiply that by 400 degrees, which is the number of degrees the mechanic is suddenly explaining, as you clutch your singed hand. By the time you reach the hose or sink or water fountain of the garage you've come up with the number 200,000,000,000, and you are trying to figure out how much a cumulative 200,000,000,000 degrees, spread over the Earth's surface at any given time, *and also divided into the time during the day these vehicles are operational, divided from 24 hours*, increases the daily temperature worldwide.

Hmmm...

That's a difficult proposition, especially under duress. The most plausible answer to your self-query, then, is that the result of such increases is the difference in temperature between the cities and the rural areas. We'll say the difference is eight or 10 or more degrees, and assume that you have experienced the effect first-hand, on a picnic in the country or some such thing. (It wasn't difficult at all, as your burned hand has already made you suspect of the CO2 theory!)

Later, as you heal, you realize that the CO2, if it is wholly to blame, doesn't seem to be spreading out as a gas might be wont to do, evenly distributing its heat retaining layers, and thereby heat, over the rural areas to the extent that it does in the cities.

The ambient temperature in cities- not for industry which is indeed responsible for the creation of *other* machines which generate elevated temperature- is increased significantly *because thousands of cars in these cities are acting as traveling heating elements*, primarily. It becomes logical, barring a solution involving cars, perhaps unused cement could be pulled from urban environments!

Continuing...

A scientist who gets a call to explain to the local news crew what he or she fears is the cause of global warming, will not, though smart enough perhaps to be able, tell the people that "cars are hot," or "cars and other industry make heat." That's too simple. The answer must be something vaguely complex, pertinent to chemicals and the greenhouse effect, something which seems esoteric, only they can solve.

It's really my opinion that these folks rarely even consider ambient temperature as a cause of global warming. Is that good news which I just told you? Well, maybe. You have a name for the phenomenon now and that is "ambient heat." I invented that.

Global Warming, "**deforestation**."

Were the cause of global warming expressly that of CO2 and the Sun's warmth, then the decrease in CO2 emissions and other greenhouse gases would lead to a decrease in the rate of global warming. By creating cars with lower emissions, enforced by law, we have attempted to lower such levels.

The only benefits of elevated levels of CO_2 are those claimed by plants which use the gas as animals do Oxygen. The gas is their lifeblood. To a small degree, it has been noted by scientists that certain areas of wild land, particularly those that are wet, have increased in mass and density for no apparent reason other than that increased levels of CO_2 on the planet have increased their growth. *Plant life on earth eats CO_2 quite well, in any event.* That odd occurrence is good, of course, for the fact that we humans kill so much raw natural landscape every year, the wild lands could use the help. Also, plants give off Oxygen, which is good for humans. They are a natural cooling element for the ground and both food and shade for the animals, *(and humans).* Their roots maintain healthy soil, as opposed to... dried up sand.

It appears, however, that the benefits of such increased levels of CO_2 are not so significant, and that is *because of the reciprocal,* this section: *deforestation*. When compared with the potential health of such systems, were it that most cars on the planet are powered by hydrogen or nitrogen, for instance, we might also be so intelligent to mind the amount of deforestation we commit.

(Note: The reader will require an understanding of "mass displacement", as introduced in this work, to find an appropriate balance of land- based mass within nature's ability to compensate for it; such balance is to be achieved specifically, without adverse effect to life on Earth as a result of the goal of this chapter, reduction in heat created by man, and without electric vehicles in place of those mentioned above, powered by hydrogen or nitrogen, etc.)

Therefore reforestation, as prescribed by those who can, to those who can do, of area-localized indigenous species of flora, planetwide, becomes an essential point on which humanity may focus. It is, indeed, the third and final chief component involved in the process of global warming.

If the above 3 items, fossil fuels and the resultant **greenhouse gases** and **ambient heat**, and the **deforestation** are now known, yet two, greenhouse gases and deforestation, are those *commonly known*, well enough. Not so, until now, is ambient heat available to be *commonly known* that I am aware. In any case, all three are herein announced as mandatory for the world's schools to understand and teach.

You and I are now two for two. *I will sign off this section with the point made, as it has been elsewhere in my work, that the negation of ice and snow at the poles is not reason to stall work on global warming. Work to make the Earth's atmosphere amenable to the creation of snow and ice on a yearly basis, as it is accustomed, and work to normalize weather patterns worldwide, etc., is your remit. Do not stop, for the product of such labor will be successful to the point that a reversal from ecological misfortune shall happen. An effect for your labor, beneficial to all in the long run- and those involved in it as part of what is envisioned to be a productive world economy- is the goal. There are no obstacles to attaining the goal* ***if simultaneous attention is paid to each of the outline components as work commences.***

I will acknowledge that it's a question of those enamored with the possibility that Global Warming is a fallacy that resultant weather patterns on Earth, due to encroachment by humanity, cannot be warming if certain "cold" registers on Earth are the result. This *is* a point that is rectified herein, by this work, and one I will suffer now, briefly. As you might expect to read, attention to the outline components detailed in this writing already solves them.

Therefore, briefly, excess cold on Earth, localized by both timing and placement, is due to 3 ecological situations, all man made, as all the maladies herein are accused of being. 1. The melting of the world's biggest ice cube, at the North Pole, displaces water, certainly, but cold water. 2. The actual, increasingly concentrated chemical amount of C02 in lieu of other atmospheric components (obviously, we consider now, lessened indigenous atmospheric gases) *equals to relative cold* as there is a decrease in temperature equal to the increase of the compressed element in the atmosphere. 3. The slow and continued depletion of the natural atmosphere is a cause. This last potentially the worst, for the whole "proximity to space" thing- simple to rectify with a little human spirit, work, and ingenuity.

With respect to ingenuity, attempting a solution to temporal and geographic cooling on Earth without using the outline of information provided will not be effective. The scholar whose work takes them to the exact chemical repercussions of excess greenhouse gas releases on Earth after years of research and an impetus afterward, for instance, to counter this address will NOT account for the state of heat itself- that which we create because of industrialization and transportation. The fact is that the innate "state" of factitious heat (or trapped heat) may not be applied thereby to chemical research properly *and it must be*. We have indeed covered this point already, as it relates particularly to the *acceptance* of ambient heat as a reality.

The definition of heat, the nature of it as an expansive, excitatory, and expulsive or repulsive agent (scientifically), is what is most required to keep in mind as we progress. It is the applied human response to the emission of "higher temperature" or *ambient heat* in general that must be applied.

Please think of it this way: By the increased pressure of an atmospheric system that is affected by heat alone, without specific gases at all as items in the equation, you have what can only be described as an advance of an innately "expansive" state of temperature, outward, an affect that is, as said, "excitatory", upon *any* gases encountered by it (with applicable results), and a "push" as it is, in general, not by other invasive greenhouse gases, but by the "front" or "aggrandizement" of factitious (man-made) temperature increases upon atmospheric elements and systems.

THE HOLE IN THE OZONE LAYER, or ATMOSPHERIC DEPLETION

There are various levels of atmosphere above us which separate us from the freezing nothingness of space. One of them is called the Ozone layer, because it is made of Ozone, and its purpose for life on this planet is to block portions of UV (or ultraviolet) radiation from hitting the Earth and its residents. These residents evolved under the Ozone layer for millions of years, and therefore have no natural, inherent ability to defend against that radiation which the Ozone layer repels. This type of radiation is not good for plants, animals or people.

For thousands of years, prior to Mans' domination of the planet, and specifically, the industrial revolution(s) on the planet, this type of radiated energy (from the sun) was kept at bay. Now, after less than a hundred years of plastic manufacturing and disposal, and the dispersal of compressed hygiene related products, humans have released a great amount of chloral floral carbons into the atmosphere. These CFCs, as they are called, are human engineered chemicals that eat away at this level of atmosphere, the Ozone layer.

Though methane is accurately considered a greenhouse gas, considering the destruction of the Ozone layer it is also relevant. Farmed cows excrete a massive amount of Methane gas. Also, due to global warming, permafrost at the arctic circle(s) and glaciers, under normal circumstances resistant to melting year-round, are melting and releasing a store of methane which has been trapped beneath them since the last ice age.

Now, as in this very moment, there is a hole in this layer of atmosphere, the ozone layer located near the South Pole of the world. This *and a general degradation of the atmosphere everywhere* allow an increasing excess of UV rays to reach Earth.

Humans located beneath where the depletion of the Ozone layer is most distinct are getting sick more than others because they are being affected by harmful Sun radiation that ordinarily, ecologically speaking, does not reach the Earth.

Generally, the increase in UV radiation that reaches Earth is significant, now, worldwide, and enough so that this problem is also. The Ozone layer has "thinned". This reveals itself in the United States as it does elsewhere via increased occurrences of skin cancer. Also, less well known- where no chemical or biological contaminants are present in the natural environment- there are found to be deformities in certain animal species based on excessive UV exposure. Analyzed for genetic alteration, amphibians, specifically frogs or toads, are affected, as well as cows themselves and other animals.

The remit to us, as people, is known in this field. This item as it relates to the enclosed threat outline is very well documented. How it may be considered a heat element, which it isn't always considered to be, bears explanation, but the need for a reduction in the damaging effects of UV exposure as it relates to our damaged atmosphere and the damaged inhabitants of Earth, is known.

Therefore, without altering the course, we shall continue banning cfc filled products and/or sanctioning companies that use them. The need is obvious, but the voice is not heard calling for *the* ban of meat. In lieu of such, until we are uniformly courageous as a species, it is deemed essential a ban on meat that is not from organically, ethically maintained cows. This is for the reason of lowering the number of animals farmed and thereby the multiple of animals creating the problem of methane in the atmosphere, but also for the general health of humans, the animals and the Earth.

Similarly, a less heralded solution which is obvious, involving shielding or shade, to capture and/ or decommission excretion from cows, their methane outbursts, but also keep the animals from direct

UV exposure is essential at this point in history. We say “capture” because methane as an energy source is a reality.

To capture methane excretions from cows to be used as an *energy source* while keeping the animals healthy is one of the greatest, *ethical opportunities* we have for the reason of keeping the cows alive for the duration of their natural lives with a profitable commodity the result.

These steps are vital to stopping the destruction of the protective layer of gases in our atmosphere and allow us to transition thereby to a recreation of it.

At the point whereby our ethical development meets with the standards of the preceding paragraphs, we may recreate the entire Ozone layer, without working against ourselves. However, this doesn’t mean we will not begin this task now. This requires replacement of other gases, as well. If we must battle something, or be against something, the eviction of existent non-indigenous gases from certain elevations, C02 and others, is our remit. We do this, starting in specific, triangulated locations, using as *geographic points* the placement of urban population centers, the track of the sun, and the entry point of direct sun at the appropriate level of atmosphere.

If, as a united planet, we fund, research and implement the replenishment of the exact original makeup of chemicals that occur naturally at that level of atmosphere, *and the levels of atmosphere which surround it,* we will be successful.

These chemicals and/or generators of such may be manufactured. *The generators created are to be lifted by balloons into the atmosphere using drone technology, to restore the entire atmosphere to that of natural effectiveness.* They may also*, less preferably to this writer, but essentially,* be stored in tanks at airports, and added in small quantities to supplemental storage tanks on the undersides of passenger planes.

Ozone itself is a difficult gas to store, however. In any case, that this will be directly positive to our situation cannot be denied, or the opportunity to try, at least, not denied. Application of the gas, it must be done.

The use of balloons is best, or primary for their ability. The use of planes is secondary, and may use some more explanation, therefore. These planes, *I'm told, testing now and flying about*, are or will likely be equipped with *one more* switch than they usually have. The activation of that switch opens storage tanks at altitude, releasing the appropriate chemical into the atmosphere where it rises and presumably affixes itself to, or mixes with, the existent chemicals surrounding the Ozone layer, and then, the Ozone layer itself.

Ozone is comfortable in its natural habitat, above us, and very fragile otherwise. Recreation of surrounding levels of the atmosphere creates a home for Ozone that is acceptable, to begin. Obviously, replenishment of naturally elevated Ozone itself is good. Mentioned earlier, decreased concentration of CO_2 in favor of naturally occurring gases and proper separation from space (to the extent that a recreated atmosphere may provide it) become other *immediately realized* positive results of such work.

This may be done in a similar time frame, to effectiveness, and with the bonus to humanity that we know for the first time exactly what we are doing. It is in relief we accomplish this, for what has occurred to destroy the atmosphere was done in ignorance, the many benefits of recreating it are seen clearly.

Appealing specifically to professionals, and perhaps so I shouldn't have to mention it later, a chemical bond to Ozone and other atmospheric chemicals, those in need of restoration, that repels itself from CO_2 and/ or becomes appropriately weightless to ensure its flight upward to correct elevation (amidst its chemically similar components in the atmosphere) is something that should be researched and implemented if planes are the delivery method.

Preferably, however, are airborne balloons. They are required.

In the event of the use of planes, the gas needs to move from the dispersal altitude, and that is the problem (which is the result of scorning the dedicated solution of balloons). If the chemical bond of required elements dissipates over time, and the resultant components are proven to be innocuous to the chemicals under duress, those we are trying to replenish, then you have found your method of "bringing" the Ozone and other gases to where they should be. It is with proper altitude, and maintenance of it, however, that an easier and more specific delivery method is found, and that is what I urge you: to suspend the dispersal method appropriately, at the right altitude, and hold it.

Chemical bonds, engineered with an *expectation* of dissipation, or break, at altitude are certainly possible, but not easily attained. Those same bonds, in and of themselves, create a chemical that is required for placement, then useless or of less efficacy, or perhaps even harmful, if the bond doesn't dissipate. (I must re-state that the fragile nature of ozone as a gas makes the chemical structure of it something to suspect as “in danger” if it is subjected immediately upon release from planes to the machine’s exhaust gases or dramatic alterations in temperature outside of its atmospheric norm, specifically, the temperature of jet flame.)

Additionally, a reality that may or may not be innocuous to life, contrails, exuded by planes as they fly normally. I am a proponent of them if they are innocuous, for the possibility that they may be laid down over urban centers, triangulated again, along the track of the sun, to offer temporary solace to inhabitants below from the piercing rays of it.

(For another study, and herein, therefore, only a statement from me now: A massive blanketing of factitious or man-made cloud cover over large areas is NOT recommended for, among other things, the prospect of harmful microscopic species propagation and obvious potential harm *thereby* to indigenous species.)

One last note pertaining to Ozone as a *contact-based* hazard to human health. Excess Ozone at ground level is hazardous to the respiratory systems, skin and eyes of humans or animals.

Should efforts be successful to combat global warming using alternative fuels and the need to replenish naturally occurring atmospheric levels be ignored, then this ground-level of Ozone may be gone. This will benefit our lungs, skin and eyes. Also gone will be any "protection" from UV radiation that we get from it, if any. This is not insinuated to be an equal deflection capability of the *atmospheric* Ozone layer itself- which is said to be 29 times the unshielded amount of UV radiation- but a percentage, in likelihood, and one which is potentially *in effect* now. Thus, we see a bit more the importance of the atmosphere and its natural ability to protect us from harm, in this case, while maintaining itself *away* from us as ground-based creatures.

MASS AND ENERGY V.S. LIFE ON PLANET EARTH

To establish the actual state of the physical mass *and condition* of Earth- as it floats around the Sun, balanced by its own weight, rotation and momentum, and the magnetic fields of our solar system- we divide the problem into two parts: one is Mass Displacement on Earth and one is Energy Displacement on Earth.

The result of each part must then be compared with the state the Earth would be in *had humans not advanced beyond the stage of evolution involving significant changes to the Earth's original state*. From these calculations, a direct line of procedure for reestablishing the healthiest planet possible may be established.

The relative weight of the planet itself is a computation to occur as a beginning. The relative weight of the planet including permanently vacated elements of it due to humanity is second, then the weight of the moon, etc.

It is important to note that the Earth can acclimate to change by humans, to a degree, itself, in defense of itself. An example is made further on in this text involving land-based mass displacement created by humans countered by mass displacement of arctic melt into the oceans, but we may also use the example set forth in the prior section: the ground-based ozone pollution's ability to deflect UV radiation.

The order is to calculate the difference between how much physical "mass displacement" has occurred on Earth since *humans* began to create it. For ease, we zero in on a period in history that can be seen as a most significant trigger point, one which distinguishes itself as it may have affected the acceleration of mass displacement. The period known as the first Industrial revolution is a standout example, seeming at first to be the largest acceleration point for mass displacement in history, at the time. Dwarfing it, however, is the mass exodus of humanity and its machinations from the Eastern Hemisphere of Earth to the Western upon the discovery of the New World, likely the most significant in world history as it relates to land-borne mass.

Where general increases over time are decidedly anathema to life on Earth, this discovery likely benefited a cause for balance, if you will, for had the machinations of man progressed as they have without discovery and development of the new hemisphere, that balance we search for would likely have been far more difficult to achieve. This assertion is not an excuse for the over development of Earth's natural resources into land-borne items, including cars, which are now our main concern, but a reality, for the time, that may have given temporary solace to those who considered the Earth as it is, a round item floating, spinning, and under inertia.

Now we've reached a point where we must *collectively* ascertain the limits of such mass displacement. How much can the Earth likely handle given the history of itself, considering all quantifiers? That is, "handle", meaning the level established by analysis "below which" amounts of mass displacement may not alter any part of the Earth's stability to the point where survival of its inhabitants (or itself, as a whole) is threatened.

Secondly, the problem is to calculate the difference between how much "displacement of energy" has occurred on and around the planet since forever, or that same revolution. If we may concede, the exchange of energy for heat, detailed in the chapter above on global warming, leaves us with displacement of magnetic fields to analyze, their change and displacement, and how much the Earth can endure, again, given the above quantifiers. It is the recommendation here that figures related to physical energy, the requirement to move physical weight to and from locations on the surface of the planet (and off of it), be calculated amidst the calculations for "mass displacement."

For example, the removal from Earth of metallic or other elements which might for an extended period hold or transmit a natural amount of electromagnetic energy into cars (and specifically batteries) *is mass displacement*. The use of such substances- whose capability after mining from Earth to hold a higher amount of charge than they might naturally, for a *newly finite* time span, to a point of complete exhaustion- is electromagnetic *energy displacement*.

You can put the cars components back where they were with the same mining equipment you took them from, and, despite its simplicity, that *is* the directive. Recharging of these permanently exhausted natural elements is impossible, however, or extremely difficult. It is probable that this area of the Earth, one from whence such substances were taken, is considered to have a *reduced* amount of electromagnetic reactivity. This *is even though*, as the substances originally laid, they had not been combined with other electro reactive elements.

Again, our order of work begins with a determination, or best assumption, of the progression of a natural state of Earth undominated by humans, as there obviously would be one. Following this, the order to determine a state relative to mass displacement *and* to determine a state relative to energy displacement.

MASS DISPLACEMENT ON EARTH

Mass displacement on the planet may be explained in three categories. Those are weight transfer due to human engineering, weight transfer due to the impact force of natural and foreign objects, and weight transfer due to melt caused by global warming. Each must be individually scrutinized for amounts, and then recombined for further study, and ultimately combined again with results from the study of the other outline elements described herein.

As it relates to Mass displacement, we are looking at the following: The Earth is floating around the Sun. (1.) Your car weighs what? (2.) That asteroid *hitting us* did what? (3.) Melting ice caps *weigh* and/or *have an affect value* of what? (Only after this does the question become: what is the "charge" differential based on the movement and manufacturing of the elements?)

MASS DISPLACEMENT SUB-CATEGORY ONE: WEIGHT DISPLACEMENT

The Industrial Revolution, from the mid 1700's to the mid 1800's was the period when society began to advance by, we'll say, "making lots of things." For the first time, machines began to be constructed to manufacture goods of all kinds. These things, these goods, were all made in a place where their component pieces and materials had been transferred to, *from* the places where they had been dug up or cut down. Upon completion, the manufactured items were then transported to entirely different locations, *etc*.

Toward the end of the period, the revolution didn't end. It grew. In 1869 the American transcontinental railroad was finished, and coal for steel, iron ore, and steel itself from the Eastern mines was now transferrable, with newfound ease, to the West, often thousands of miles away. In Germany, as further example, the Ruhr Valley was just beginning to produce what would equal and occasionally surpass the yields of the coal mines in the U.S, Western Europe and Russia. This yield was also transported, often thousands of miles, to other nations entirely, in the form of automobiles, heavy munitions, ships and manufacturing elements.

This era represented an unprecedented, literally exponential increase of factitious mass displacement. It was, in retrospect, greater displacement of land on Earth than even Earth itself had accomplished save for in one instance since prior to the previous ice age: The cumulative separation of the continents due to continental drift!

It can only be assumed, at this very point in human history, around the time of the development of cars *which ran on gasoline*, and whose motors were thereby hot, that something strange began to happen: *The primary mass displacement on land, the actual weight moved by humanity, began to be equaled, or surpassed, yearly, in what can be considered "secondary mass displacement", that of excess water run-off from the poles due to global warming. This phenomenon has its own section, as noted, but bears mention now as we are at a point in time which witnesses it.*

In any case, *calculations to* quantify amounts in this short period can be borne *easily*, but only when using the denominator "millions of tons." Since that time, it is “billions of tons.” This is as much so because the figures may not be specifically available *yet*, as it is for the sheer amount of weight under scrutiny. The sooner we begin computations on weight transference to date, the sooner a system may be developed to monitor it. That is to say, the difficult proposition of "doing the math" becomes moot, and the task of repair, accessible.

We see the need to disseminate the problem of historical data accumulation to the world rather than do otherwise, as we have historically- kept the need and science esoteric or ignored entirely. The multiplicity of resources given a global curriculum makes results inevitable, comparable and sharable.

For its obvious relevance, we may look at the *car industry*, first, in North America. Considering the cars made from steel created by iron ore that is mined from *specific locations*. That "weight" is then turned mobile, as it were, and unleashed upon the surface of the planet again, *in specific locations*.

This is not done evenly, of course, but rather, in favor of urban centers, which thereby become the mass displacement element we are led to study concurrently: *the weight of cities*.

Some urban centers are bigger than others. Southern California and New York are good examples. Each of these *areas* weighs millions of tons more than it did 150 years ago. The areas from where these cities got their steel for both cars and buildings *and their cement* weigh millions of tons less than they did 150 years ago. They do so, however, again, *specifically*.

We may go back, as a reminder, and mention that the Earth spent hundreds of millions of years creating the natural state that it had prior to the industrial revolution, (*and prior to the construction of the Egyptian pyramids and the great wall of China.)*

On a national level, like the example given, the problem of displacement is localized, and when initially viewed on an Earth- wide level, as if by an astronaut, it seems not so shocking. Worse, perhaps, *than domestic transfer of weight* is the problem of international weight transfer and deposit.

How many more German cars are on American soil than American cars on German soil? What percentage of Japanese cars is made from Japanese steel? Chinese? How much of what was once "over there" is "now here," and vice- versa, etc.

The major categories for factitious mass displacement are oil wells, mines, cars, cities, impact force of bombs, deforestation and displacement of water from the poles due to global warming.

Deforestation, if it may be touched on here, seems small, relatively speaking. However, it represents a mass displacement issue and an atmospheric issue, as well. Oil wells do, too. Wood and oil and other elements have been *burned*, and a specific tonnage of land-based mass has become a tonnage of emitted CO2 and burn gases- *newly created atmospheric gas*, which is suspended, rather than a contact-based mass of Earth, and therefore only a burn-resultant *mass* remains, and that shall be calculated for both wood and oil.

MASS DISPLACEMENT SUB-CATEGORY TWO: IMPACT FORCE

In 1944, after creating hundreds of military tanks per month, and storing its supply of them for nearly two years, Russian forces headed south toward Germany. Each tank weighed a certain amount which is relevant to the issue of *weight which we have covered above.* Each of these thousands of tanks fired a big shell that landed somewhere between its own land and Germany, and that amount may be calculated. German tanks went to meet them, and American armaments headed northward from the South. We will speculate erroneously here that on the trail of war, fifty thousand tank shells struck the Earth with an impact of 650 feet per second, at 12 pounds per shell within a certain period. 253, 500, 000, 000 foot/pounds of energy hit the Earth. This excludes the force of the explosive charge of each shell. We come to our conclusion again using the formula E= MC2 where mass is the 50,000 tank shells multiplied by their weight of 12 pounds plus their explosive force effect, multiplied by the constant squared, 650 feet per second...

It is decided that the force was directed at the Earth at a 35-degree angle of impact, average, toward the Earth in a *certain variance of direction,* South by Southeast, North, etc.

On the other side of the world, a baseball player hit a home run that landed on a soft piece of South American dirt with a force of one baseball times 100 feet per second times ten ounces, or whatever a baseball's weight. A disparity becomes clear, and it is one which needs to be deciphered.

The calculations, cross referenced with the weight of the entire planet, the hold of its magnetic field assisting its positioning in space, its actual positioning in space and its velocity, tell us what was done by impact force during the specific time in question, and whether or not the effects would be permanently damaging to the cycle of the Earth, were you and I not here to do something about it. The impact of flying items is what we speak of now, not just the weight of items.

This amount of weight, landing during this barrage of wartime fighting, is small in comparison with the total amount of surface impact unleashed on the planet. Nuclear bombs come to mind. That's quite a bit of impact force, wouldn't you say? Your Earth is therefore quite strong, the lesson.

It has been suggested to this writer a figure I'm not certain of, but generally, that trillions of foot/pounds of force were applied by bombs to Iraq in 1991, during the 1st gulf conflict.

It seems quite impossible, given the many impacts that must be considered, to calculate whether we have doomed ourselves by our wartime efforts, occasional plane crashes and building collapses. Add to that the movement of weight, in general, as mentioned in Sub-Category One, there is quite a mathematical problem. It is one that is solvable by research and calculation of historical events. For this reason, however, (its difficulty) it is of most importance to first establish, also, as mentioned, the threshold of forced weight displacement on Earth, by pinpointing its normal qualities, those that existed for thousands of years during normal Earth cycles- and then decide the strength of force necessary to upset that cycle. This has been done, again, I feel certain.

In any case, the figures decided upon, given the calculations necessary to get them, should be, I would suggest, compared with the figures of maximum weight displacement/ change that occurred at the most dramatic natural moments in Earth's history. These include an Ice Age, or meteor, or successive nuclear tests. They may also include a cumulative melt from the poles based on current effects of global warming, which is the opposite, one might suppose, of an Ice Age.

It is my fear that the force found necessary to create a dooming imbalance will turn out to be quite minimal. It is my assumption, however, as said, that this is not the case. It is *the* hope that the destruction and force inflicted on the planet has been done to a less than lethal amount of pressure from impact force, and that the Earth is therefore not in danger as we add these variables to the totals of mass displacement attributed to the works of man. The next sub-category, concerning Global Warming, addresses this issue of hope, somewhat.

Mother nature is your friend...

MASS DISPLACEMENT SUB-CATEGORY THREE: GLOBAL WARMING

The process of calculating what we call the "Earth's trajectory" under normal circumstances, those not influenced by human behavior, is sufficient to develop a condition *or* position (actual physical position) of the planet over time. Unfortunately, there is no constant for the potential damage done to this trajectory during the past 150 years, or "ever", (to the extent that it makes one feel better considering it.)

We don't want an "assumption of trajectory" that is flawed, for the fact that we believe, based on reverse calculating, that the Earth's trajectory arrived at what it did via *a constant* trend away from its norm. It is more likely that a constant trend away from a norm was then altered which pushed the trend in a different direction, one that a single "bearing", if you will, cannot decipher. This is why an understanding of all *major events* pertinent to mass displacement (and energy displacement) is relevant to our study, and why the actual condition of what would be considered natural needs to be determined.

To explain, we can use the example of a free- floating ship on the water. Its original course was set to take it to an island. It was then found to be headed toward the middle of the ocean and stopped. If we calculate a new course based on only the ship's position in the sea and the new required course in a sea whose tide is altered via the tide or shipping lane currents, then we stand to fail a second time. That is, our new calculation to send the ship to its destination, including the force of sea current *originally calculated* or *the force we assume for the ship's new location* would be moot for the *actual, "current"* sea current direction at the time of calculation.

Our Earth, without an easier way to explain it, is fat in the middle, now, and light on the top. A mass of weight has transferred, melted ice, to water, a *safety mechanism* the Earth has for her own survival, enacted to balance her own mass displacement, to compensate for factitious or man-made mass displacement on land. The byproduct of factitious mass displacement, the burning of fossil fuels, and the other machinations of man, *is heat.* That is what causes the phenomenon.

A second phenomenon exists potentially, relevant to permanent evaporation of water, made easier by a decreased salinity of the oceans, now diluted by the run-off, and increased solar electromagnetic assault. This second phenomenon represents only one potential course alteration, as an example of what we have just discussed.

A reduction of mass displacement on land is called for, therefore, as we are close to lacking such a counterbalance at the North pole. The reason for this is that the ice cap is almost gone.

Obviously, a reduction in such mass results in less displacement of equivalent polar mass, because the resultant *ambient heat* is lessened.

We see now an opportunity to use each of this item's outline components in one example, and we have yet to reach the section on Energy displacement. If less people (overpopulation) drive smaller cars (mass displacement), emitting less heat, creating fewer greenhouse gases and needing less room to maneuver between the trees (3 tenets of global warming), then less loss of Ozone and other necessary atmospheric gases occurs (Ozone/Atmospheric depletion), resulting in less electromagnetic assault on Earth (electromagnetic energy displacement).

To accomplish the initial goal relating to mass displacement, however, specific weight limits on production cars are required to be enforced. The application of such enforcement worldwide is recommended.

Unfortunately, it is considered a scientific secret, though the terminology used here may denote something new for you. Those few "in the know" consider as a mollifier, a calmer, the potentially positive effects of second-hand factitious longitudinal mass displacement by displacement of water. However, the effect of the polar runoff equating to humanity's misdeeds on land is as dire a warning as we have for the depletion of the poles, mentioned.

Zero effort on behalf of the world community to engage in an open discourse to identify it as a scientific reality has occurred. When such effort is put forth, it shall be known already that *Latitudinal and Altitudinal mass displacement*, which are affected by the same progressive, factitious displacement of mass, are distinct subject matter which present their own quantifiable effect on the situation and the solution.

Mass displacement because of *melt* from *Global Warming* affects our situation here in much the same way as the effects of “weight” displacement explained in Sub-Category One, above. The difference is that- though permanently "lighter" at the poles from where increased water runoff and decreased water accumulation and re-freezing occur yearly- we *see mother nature’s attempt* to lessen the effects of mass depletion in areas affected by mining and deforestation.

Curiously, no other element might have been used besides fossil fuels to accomplish this heat to mass equation. It has been suggested in this text that the mining of materials for electric cars, specifically, and the exacerbated depletion of charge of such components is a rather terrifying scientific reality. Now we see, again, that if electric cars were the *original* choice to power humans on the roads of Earth rather than gasoline, a rock might just be writing this for carbon pebbles to read as we would be extinct already.

Nevertheless, despite the very modest accommodation to gasoline as an original fuel source, alternative fuels, albeit non-electro reactive, *are* our current remit, as a species, provided the other outline components contained herein are attended to.

Please keep in mind, the amount of factitious heat created by man without the use of gasoline is distinct. It is to find the actual ecological balance we require without dogmatism of any kind that is the requirement. Only then will the actual land-based fuel sources which are required be known.

The exasperated flow of water from the North pole has eclipsed runoff from the Antarctic continent, and it is no coincidence that the mass displacement (and heat) created in the Northern hemisphere of Earth exceeds that which has occurred in the South. Our first point of reference for the future is this one as it leads us to the example of Mother Nature's inherent ability to accommodate herself in support of humanity's growth, though this may only be so for a short time. The northern hemisphere has a mass displacement issue on land which is excessive, relative to any other land-based example, and is corresponded to, specifically, by *the effect of water run-off for which we are hypothetically, briefly, fortunate.*

Water is "recyclable." Flood the mines. *A concerted effort by the residents of Earth to do the simple things required to make the situation tenable represents a possibility for Earth, its inhabitants, and, as an aside, the economies which they foster.*

The fact that flooding already occurs in places which have been depleted of resource weight means problems relating to mass displacement have compensated themselves somewhat, (as above), but only so much so the situation may be used as an example of *what we need to do* with excess runoff water yearly.

Water accumulated by our efforts, on land, may be maintained to our benefit in consideration of the resource weight we intend to replace, but also to the counter of, or response to, the effect of accelerated evaporation or the needs of a growing population.

On the subject, *desalinization* of ocean water is an interesting science, and may be a process that is one day required. Increased run-off from the poles equates to a reduced salinity of the oceans, and desalinization to the creation of potable water for drinking or irrigation is a means whereby humans benefit by helping to restore the normal balance of salinity to the oceans as a byproduct of the labor. The salinity of coastal regions' *freshwater* has increased, otherwise, because of sea level intrusion due to oceanic rise, unfortunately, and it might be nice to realize some benefit from the situation.

It is important to note that efforts to desalinate ocean water for freshwater must be complete in their result, as historical misfortunes based on remnant, undetectable salinity have occurred. This is "oligohaline" water, in English, and it contains a tasteless measure of salinity between fresh water and salt water.

I reiterate to you that the Earth is not without her tricks, that mass displacement to the oceans just as well mollifies or eases the effect on the Earth that would be attained had man only their own, physical, non-water weight to move about. This situation of protection is temporary.

The presumption here is that this situation is capable of being affected by the positive works of man. One such type of work, not yet mentioned here, is localized movement of land-based *mass*, not water, to areas suddenly, ecologically speaking, devoid of it. Recycling centers, of a size unrivaled in history, located where mines are, internationally, for instance, is surmised to be the directive. (Examples of the specific, currently displaced resource weight items, or "things that weigh the most," most of which are recyclable: Iron, steel, aluminum, cement, wood, etc.)

This is an attempt to solve two problems by one solution: the solution of movement of weight to counter problems of both mass displacement and loss of natural resources on land. There is another, however, and the third deals with the specificity involved as it relates to electromagnetic charge. An issue not to be ignored during such work is that charge retention of areas where iron ore existed, *for instance*, may not, with replacement *by non- metallic elements*, match the original electromagnetic charge retention ability of the *natural Earth* and, therefore, the option of reprocessing and recycling only *indigenous* materials at the site from which they originated seems a requirement.

Electromagnetic Energy Displacement is covered more thoroughly in the section which follows. It will be, for *altitudinal mass displacement*, for example, to use “altitude from center” and “height of mass in relation to actual to gravity”- as starting measurements to begin analysis. This is how a proper accounting of mass and weight on a rough circumference by incremental gradient skywards becomes possible. “How might a car affect the situation of mass displacement on Earth at various altitudes?” *We will find the analysis of Energy Displacement to be similar, no doubt.*

This section may be closed now as indicative of the need we have to account, collectively, for the mass displacement we create on Earth. That situation is vastly different as a result of human encroachment on the surface of the planet than what it might have been without our assistance. You’ll please remember the psychological blocks mentioned earlier in this address. Remember, please, how they might have played a part for some decades in the assumption by the masses that the greenhouse effect was most important to consider as contributory to global warming- rather than the point that actual or ambient heat created by man is as much or more so to blame for it, (as well as man’s destruction of natural wildlands.) Looking into implications of an imminent “polar flip” has an electromagnetic implication to blame for the occurrence, and an electromagnetic polarity reversal as a potential result, oftentimes hypothesized by my inferiors to a result of something other than a physical change in the Earth’s normal position.

I might suggest (without the implication that I believe it is likely) if one were to occur it would have nothing to do with the esoteric of a naturally occurring polar-magnetic situation. It would have everything to do with the obvious, yet again: *human application of electromagnetics* and, as much so, the actual factitious *displacement of mass* on Earth beyond the parameters which allow Earth to sustain its rotational and gravitational norm.

ENERGY DISPLACEMENT ON EARTH

Weight as an evaluation through gravity, movement of it or “mass displacement” on land, and force impact from bombs, etc. may affect the movement of this planet and its momentum and/or directional “energy” relevant if we choose to think of it that way. This section, specifically, deals with the Earth's own charge, magnetism, polarity and electromagnetic emissions.

We call the Earth’s charge, magnetism and polarity the “Earth’s own” to differentiate it from resultant energy (and temperature) of other celestial items (the Sun in particular) and the electromagnetic emissions of mankind. Weather energy is, to some degree, an effect of the above, a *result.* Significant here is that weather patterns may be a *prelude* and indicator of a more severe natural calamity that is inbound for the change in the Earth's Energy and Mass Displacement, Global Warming, or the depletion of the Ozone layer, etc.

Have you ever heard the theory that, since it is similar in "shape" to an atom, the solar system may in fact be an atom in some incredibly large Universe? Since we know the size of atoms, we can calculate the size of that other Universe. That theory suggests that atoms surrounding us may have little societies developing on little electrons that are planets, so small that we could never, for their scarcity amidst the billions of atoms, specifically, have any chance of finding and then seeing them.

I will confide in you, this thinking is what fuels many quantum scientists' Time Displacement theories, (for our solar system simply must move slower than an atom.) I'm... not a fan of most quantumists to speak of. Theirs, minus relevant, kinetic ability, is an irrelevant pursuit, at best, or potentially disastrous.

The only reason the theory is referenced now is because the solar system, like an atom, has its own electrical charge. That electrical charge is directly associated with that draw or push (magnetism and polarity) of and between its components, *and the result of such, specifically*. Those components are this Earth, its Moon, the other planets, and the Sun. Pertinent to this Earth, the magnetism and polarity have to do with the bond between the Moon and Earth, and therefore the tides *of Earth's oceans*; the Sun and the Earth, and therefore (resultant, again) the planetary temperature and lighting; the Earth and its atmosphere, and the Earth and its own gravity, etc. Space debris is another component group, (the effect of which may indeed be like that of the moon and the tides of Earth if the mass of satellites we put up over us becomes much greater.)

If we spin a dusty ball, a soccer ball or basketball, on our finger for five minutes, the ball will no longer be dusty. The dust, due to the ball's momentum and the force of the air around it, gets spun off the ball. The Earth has been spinning around its own axis, like that ball, and around the sun, for millions of years, yet obviously it has not shaken its surface material off of itself. It has, as a combination of its mass and charge, better explained by your local physics teacher, created a circumstance of *gravity* within and outside its atmosphere, the pull of which is lessened the farther away from Earth the measurement is taken.

Gravity occurs as a natural attraction polarization between the Earth's core, based on its mass, and all the items and area surrounding the Earth. It is a force, if you will, that affects all things regardless of their metal composition. Its strength of attraction is based on the "mass" of the objects in question. This force, on Earth, is very strong, yet it is constant enough that fragile eco-systems may be developed and maintained within its field, and atmospheres may be developed at specific and *relatively* unchanging altitudes from Earth. The force is so strong and so constant that the only change we *generally* "sense" in the gravitational pull of Earth is the shifting of the tides of the ocean. This is due to the Moon's relationship with the Earth, specifically- the Moon being the largest single mass in our planet's field of gravity, and therefore, the holder of the second strongest gravitational pull within that specific rotational field of influence. The relationship between the Earth and the Moon, and that satellite's movement around the Earth, creates a regular change in polarity which is reflected to extremes daily on the shores of the three major oceans on the planet.

To this point, before an explanation of the change of charge energy on Earth: The single best thing humanity can *do if the only extinction level threat to Earth is a loss of gravity* is to restore its fixed mass. We may begin with a gradual cessation of uranium mining. This way we can also do away with the use of uranium for the augmentation of conventional munitions.

In the year 1882, Edison Electric in New York City began to distribute electrical power. That was 1,882 years after the beginning of the western Christian calendar, 33,400 or so years after aboriginal Australians scrawled the earliest known paintings on the walls of caves in Queensland, Australia and, probably, a full 50- 60 thousand years after humans *first* left the continent of Africa. In the short time *since* 1882 every person in the civilized world has gained the use of electricity.

The possibility is that the Earth's magnetism, polarity and perhaps even its gravity has changed as a result. It has done so *independently* of any effect of such changes due to displacement (and conducive depletion) of otherwise naturally occurring electromagnetic materials which comprise it.

Could the relationship between the Earth and the Moon be affected because so much electricity is used on the planet? Could the Earth's natural charge be polarized negatively or positively for the fact that electrical current is now and constantly being generated on the surface of the planet? Our premise states in response to each point that it is.

There is a theorized link between the rise in leukemia and exposure to electricity; people living under electric transformers, perhaps, developing the disease more frequently than others. Well, if leukemia or other serious health condition is related to exposure to electricity, then the possibility that such is a microcosm of the effect of that same electricity on the state of the Earth is one that cannot be ignored. The Earth *is* a giant biosystem. It is as well considered alive as anything, and the greatest association between it and you, barring an electric shock, this example.

Again, as with the evaluation of Mass Displacement and its results, the change in polarity between that level which the Earth is accustomed to, beginning epochs ago, and the level existent now, developed with the sudden assistance of its own most studious inhabitants, is what is important.

If the Earth had an electrical charge of x for centuries and then suddenly became subjected to charge of x plus 4, it might be "shocked". If the level, however, grew only slowly from that level of x to a level of x plus 4, due to its own evolution- created minus the conscious “intent” of its inhabitants- then the change might not be so dramatic; because it developed such a charge on its own, it would be considered a *natural charge*.

As said, 150 years is not a long time for such a change in surface rates of electrical activity, or “currency,” to occur. Therefore, the problem of dealing with excess electricity (electromagnetic radiation) becomes necessary and does so at some very specific rate and scale.

To another point, emissions from Earth affect the surface of the Sun as its emissions do so themselves the surface of Earth. The Sun is, thereby, very obviously, the gentlest and most sensitive of giants we will ever know. One may safely anticipate the day we break through from the active monitoring of "solar flares," for example, to the *corresponding measurement of out- bound energy from Earth* which we may use to monitor our potentially irritating affects upon the Sun itself.

If retention by the Earth itself, its physical mass, of factitious electromagnetics is a concern, then refraction to avoid an increase in emanate electricity is the initial requirement. Encasement may be required of both mobile and stationary sources. The need for refraction to avoid enraging the Sun begs attention. Measurements of wave severity are what matter, and they come first. In order to keep the most intelligent and occasionally famous scientists on our planet happy (general scholars and educators, too), those men and women to whom I hope this address goes, from being enraged by the omission: your planet is very much iron, very much metal. Its tolerances are calculable, yes, but rather immense as well.

As it is, minus the affect of electromagnetic energy on Earth, there is only (in a similar category) the affect of raw magnetics. A point is made now concerning the mining, manufacture, use and discarding of items which are (inherently or made to be) magnetic- without charge or discharge of any powered electromagnetics or charge emissions. Inherently magnetic or potentially magnetic materials buried under ground are "shielded" by the Earth itself, placed and situated after millions of years of the planet's development. Discarded items, from consumer electronics to medical and industrial applied materials, are not. This represents a shift in the points of severity of "inherent magnetic force" on Earth, one that relates kinetically, not potentially or even somewhat debatably, to the polar balance of the planet. Some is missing, some has arrived, some depleted, some increased, etc.

Hypothetically, the outline of information explained in this document precludes threats related to this situation by taking the items as mass, and what may be called "unpowered" electromagnetic energy discharge from Earth (in addition to the point of thereby altered gravity and, again thereby, altered mass displacement.) The point, based on "mass" alone, is severe enough to indicate to humanity that these materials do influence the situation of life on Earth, and should therefore be measured and acted upon by humanity if required. That they are specifically, markedly, magnetic by nature, or measured according to the scale of magnetic qualities (importantly) yields the following: that specific placement of such items as the result of our efforts to measure the affects needed to right the situation faced by Earth makes them, to greater or lesser degrees, important pieces of the restoration puzzle of this planet.

If it is not known to all mankind that the distance between the Earth and the Moon, or the speed at which they move, or their positioning relative to each other is changed in the last century- more so or less than should be, given the history of the two- then it is obviously not surprising. Earth's tolerances have not been met, or it is not the type of information which will benefit the public to know differently. Now is when you or somebody goes forth to find such information and decipher it.

I think to myself, I've come this far and laid forth some very simple methods by which other great planetary maladies may be dealt with. I should do so again. But I can't; I do not have the scientific acumen for it. In this case it is: "Probably electrical shielding, applied with calculated timing and effectiveness," as the best I can do. Perhaps calculated transmission is in store for our future, as well. My kneejerk reaction to the thought of it is to disagree with "the" Mars idea relevant, if nobody minds the mention of it.

The use of Mars as a base for the firing of missiles or establishment and use of magnetic emissions facilities for the point of diverting inbound meteors indicates that problems we have on Earth, likely due to Electromagnetic Displacement from Earth, as noted herein, are not being correctly attended to in the first place.

This point, not from Earth, but brought down to it, brings another. It is a sad one, but the excuses which the foolish need to justify themselves may indeed be the scholarship of others, and so it is. The deflection from Earth- not, in this case, Mars- of inbound space debris (and meteors) using *nuclear* missiles from Earth is or isn't a good idea. We should keep in mind, for the bad we perpetuate, the potential bad, we have small points of light we may indicate to critics. The ostensible use for the weapons we possess (as opposed to the conventional intent of them, to kill each other) to do another far more altruistic job is not a good enough justification for the threat we have to ourselves as a result of creating, storing, and keeping them. *Multiple kinetic force projectiles* is an idea of potential merit in lieu of the radiation.

The only other mass diversionary system possible is likely one of applied magnetics. This type of system works or would by adherence to a projectile of a charged element, one time, and subsequent use of the element as a guidance point for subsequent diversionary "payload" delivery. Similarly, applied, beam-based magnetics, are hypothetically possible to be triggered at stationary sources (say the moon or Mars) for the purpose of attraction or deflection of iron ore- based meteors, etc., but does so with a result, each time, that is equal to the effort, and not likely beneficial in the long term for Earth.

Meteors come in various sizes.

G-d invented scientists, just like He did the preachers and everyone.

Earth's ecological threats boil down to 7, and 5 are written above. Two remaining, the waste of natural resources on Earth, things we need, and the improper disposal of caustic substances on Earth, things we don't need, may be dealt with by *recycling* and *burning or storage*, respectively.

Two issues of such biological importance that they cause concern for those in the civilized world, mentioned frequently enough I have to endure it: the genetic alteration of crops and co-operative use of pesticides in the effort, and the prescription of engineered chemicals by the medical industrial complexes of the 1st world- in the form of antibiotics and neuro psycho pharmaceuticals, in particular- may be solved by *burning*, as above.

Be good to each other, men and women. Treat the elderly and children and the infirm with all care. For want of another opportunity to say so- treat animals and your floral/ plant systems well. "Don't kill it *without reason*." Animals are intelligent, compassionate creatures. They feel as you and are unable to communicate with language. You are their shepherds. If you work from this plan, you will be successful. Declines in major suffering are my forte, they're required and very possible. That's it.

Your friend, Will.

* This item is accomplished so that this planet's inhabitants may live for the longest time possible. The environmental, ecological, and astronomic threats to Earth *defined* and each *defied* by one outline, and then given to educators worldwide so that they may teach the children of Earth accordingly, meets the general requirements of that which I have been repeatedly accused: pretty good.

The opposite of such work is any action which creates for the first time in history a calculable end date for humanity which precedes the estimated time of departure we're at now. This is hard to avoid when we're fussing about with each other constantly, but it is so. Fortunately, despite strong efforts, this has not occurred by some point or action which is obviously to blame, and we should keep it so.

An example to this point, cleats in the ground used in order to collect or channel the inherent electromagnetic energy of Earth in the form of the Earth's own groundwave energy as a guidance system for the wireless transfer (between them) of electricity- that which is now created in amounts transmitted by wires, those used to power many things on Earth- is possible, and not advisable. Though this method has been mocked once or twice in popular culture, in all seriousness, and hopefully finality, it is, repeatedly, *not advisable*. An example, a banal one, as to why: a cleated car battery. A scientist whose methods are more specific than my own will rather easily, in a week or two, deduce the reduction of tenure of Earth's inhabitants from "not sure when" to "exactly when" as a result of the employment of such techniques: those which eventually, in progression, leave Earth a dormant shell.

The calculation, excluding the horrors of determining localized results, probably involves a bracketed reduction constant multiplied by area and divided by an unknown integer (the circumference of emission of the captured groundwave). Then the computation subtracts a little, and then subtracts a little more. Then there's another constant representing negligible recharge based on the sun's rays and gamma rays, followed by the rate computation of what kills us as the result: the physical imbalance of Earth in it's rotation to the point of complete destruction. Probably. If it doesn't then it's something complicated like that. It probably involves a multiplied, bracketed computation of the Earth's reduced charge ability over time...

A depletion of the Earth's electrical field over time that is *greater than* that which we create using materials which make various batteries *and* the bombardment of our planet with the resulting wave energy we create as we harness the power of those substances *is not a good idea.*

Calculated depletion of charge is not something to be taken lightly. It is therefore, if at all, to be accomplished for the result of lowering the existing *factitious* charge assault only, with no other motivators, and therefore accomplished initially as suggested in this document, by shielding. An example from this work *of* this type of work is the use of balloons to replenish the atmosphere (versus the use of planes) explained with a similar advisory; don't employ works toward extension in hopes of other, secondary returns, or as a secondary mechanism to existing technologies which are bad. In so doing there is clouded judgment- the problems inherent, used as some way to "also" accomplish something else, take from the altruism of the result, at best.

Your action in defense of yourselves is for that result only, nothing more. Results of this entire work, in this altruism, for the lone, humble effect of survival, shall reveal such "secondary bounties" in the future, and those "as results", only, not "by intent". Any intended as a precedent, secondary goal to *exclusive* action will be considered positively granular, and possibly, as I have implied, internationally treasonous in hindsight.

** Understanding Earth's 3 major states of matter as *two*, Mass and Gas (in lieu of Solid and Gas and Liquid) is a new prerogative of the scholars of Earth. In the United States, at any rate, the lesson plan to children in schools is that there are 3 states of matter, which is erroneous.

Liquid's *result* upon the Earth and its life forms is not distinguishable from the effect of its mass, or "solid" form. Fluid dynamics is only a method by which this mass moves, as all mass does, based on its physical characteristics.

Fluid dynamics do not precede, by scientific analysis of any kind, the reality that the inherent nature of liquid is as a mass, contributory to weight born upon the rest of earth in unity with all things known as "solids." Neither can separate itself from the other for gravity, molecular weight, magnetic attraction etc. Liquid exists within the same parameters thereby as solid.

Similarly, fluid dynamics do not precede the *effects* of liquid on the Earth, either. As a matter of fact, easily found citations by professionals implying vibrational differences between solid elements bound together and liquid elements, also bound together, are erroneous. H, O and H2O *each have* vibrational characteristics which are singular to themselves, but *each solid* and *each gas* has a singular vibrational characteristic, as well. These cannot therefore be understood to be different in any way than the effect of a given *solid* (or *gas*) on the Earth.

Mass, as a new term for solid *and* unevaporated liquid, precedes fluid dynamics in importance. Therefore, the scientific fact is that we are left with "MASS" and "GAS" as the two "matters". *Mass now denotes* solid matter *and* liquid, and *Gas now denotes* gases as we know them *and* the evaporated form of water, ascended water vapor.

This will be noted for future textbooks and lesson plans worldwide, and I am not at all pleased to have to inform certain scholars that more recently seized upon theories- those of a fourth state of matter sharing both the qualities of solid and liquid- are moot. "Plasma" as I believe it was called, pierced, releases a *gas*, and whole, seems a gravity bound *mass*, because *that is what the two states of matter are:* one that bonds to Earth and one that doesn't.

The discovery of what was termed plasma preceded what I now count as at least one more published attempt to create matter where there is none, an attempt at a fifth state of matter to even further complicate the path to what we seek, which is reality. How famous, I wonder, will the person be who discovers the sixth state of matter?

A lesson of similarity, for referral, is that of the "flow" of both "sand" and "water". The method *and appearance* of their movement is quite similar. This is because the movement of these items IS quite similar, and the basis for this characteristic is that both are attached to the Earth. This state lasts until particles of either sand or water become separate from the mass of themselves to become gas. *One does so* by erosion to a state so minute the wind carries it into the atmosphere as an element *and/or* so minute that the charge remnant, which is singular to the element, is repulsive enough to that of the localized Mass that it floats. *The other does so* by evaporation.

*** To *global cooling advocates*, as mentioned, you can't switch global cooling for global warming on the outline. If you do there are created ambiguities and absences of important, real affects to our situation. I understand from where "cold" minded individuals are coming from, and it might just be the cold of space which eventually does us in. However, the situation of global cooling is one of the most significant to be affected by the outline subject matter discussed herein.

**** To *global warming adherents*, a note on the most recent of seemingly dire indicators of our situation- the first of which, a polar temperature spike in the Arctic of Earth suffered in the winter of 2018 to the point of 30 degrees. As latitude shall be examined concerning Mass Displacement, as proscribed earlier, latitude may likely be blamed for Arizona and Texas surviving the coming summers.

The Arctic and the continent of Antarctica comprise the areas of Earth with the least "area by latitude." An Arctic spike in temperatures, isolated there, is simply that, isolated to the pole itself, possibly to grow, but for now, likely to correspond directly to the smallest area of latitude in the northern hemisphere: the planet's pole.

This situation intensifying may be marked, from the top down, as it were, and fortunately should affect the colder regions of Earth before the warmest. Also, as latitudes reach the equator from the poles, ever increasing areas of *dispersal* based on the increased dimensions of the

Earth by latitude itself are available for retained and factitiously created heat.

***** It is not a possibility to see an ignition of atmosphere based on the works of this plan. Hypothesizing as to what might create such a circumstance, I can't. It wouldn't be possible under normal environmental circumstance, never mind so under the current circumstance of CO_2 over-abundance in atmosphere.

We find when searching for the most combustible element to be released into air, it is not an element whose combination with Oxygen creates a large- scale ignition. This point is made because, generally, in search of a method whereby a large-scale ignition of atmosphere occurs, it is Oxygen *in unnatural combination* with other gases or combustible elements that creates the potential. These elements, those we might look for, are evaporated crude oil, refined oil, gasoline, and the gases, butane, propane, and *natural gas*. There is none which exceeds in its combination with Oxygen the general combustibility of *natural gas* that I am aware of (innately.) Accumulation of this gas in amounts which can be considered remotely relevant in an uncontained atmosphere is doubtful, to say the least.

To explain the impossibility, a purpose served by lightning itself, unexplained by scholars that I have seen, another of Earth's safety mechanisms: constant, non-stop, attempted ignition by lightning. As a melting polar region is a safety mechanism for Earth's inhabitants relevant to mass displacement, explained earlier, an unheralded gift from Earth itself, lightning has served as one since its beginning as an atmospheric occurrence.

The possibility that *any* gas or accumulated element might reach a point to be lit in combination with the gas Oxygen, or any other, is negated as *Earth has performed the regular attempt to do so*, serving as a potential pre-ignition of gases which may presume to be so bold as to endanger, by over accumulation, Earth's life forms. *As said, lightning,* usually in combination with rain, an extinguishing agent, has served for millennia to ignite (and extinguish) any potentially accumulated items on *or* suspended over Earth.

With such an ignition there is a secondary safety mechanism: a disruption of accompanying cloud cover to rainfall and the immediate "further" extinguishing of items afire. That's the way it "should" be, at any rate, though unprecedented, temporally localized drought has impacted this system.

Elements that are possible to be released into atmosphere being accounted for leave only one other potential item of concern, an item which might be able to ignite Oxygen in atmosphere more readily than lightning. By sheer duration and temperature, the only item of such value *hotter* than lightning to contact our surrounding space is a nuclear blast. This has been tried in ignorance, more than once. It is at this point we may call to notice the proximity of the hole in Ozone to such test sites as suspicious, and not much else.

****** To all... if it *is* at all possible... do not in the future force people to work on this plan or subject matter. It should bereave me greatly to find, after my death, at an old age, of natural causes, as I look down upon you from my resting place in heaven, or Valhalla, or computer camera, at the remaining of Earth's living souls, that the impetus for work on this subject matter is not "personal interest", but rather some futuristic (or historically verified) manner of slavery or servitude.

Thank you. It has been my pleasure to save you from extinction.

Sincerely,

William Charles Duffy III

Made in the USA
Monee, IL
08 November 2026

67388646R00035